CLIMATE CHANGE

Science, Impacts and Actions for a Sustainable Future

CHAPTERS

1. Introduction to Climate Change: Key Concepts and Global Context.

2. How Climate Works: From Solar Radiation to Atmospheric Dynamics.

3. The Scientific Evidence of Climate Change: Data, Models and Trends.

4. Impacts of Global Warming: Key Areas of Concern.

5. Changes in Climate Regimes: Extreme Events, Vulnerable Areas and Migrations.

6. Oceans in Danger: Acidification, Sea Level Rise and Marine Biodiversity.

7. The Role of Human Activities: Greenhouse Gas Emissions, Deforestation and Pollution

8. Energy and the Green Transition: From Fossil Fuel Dependency to Renewable Sources.

9. Sustainable Agriculture and Food: Climate Impacts and Agroecological Solutions.

10. Resilient Cities: Urban Planning, Transportation and Pollution Reduction.

11. Threatened Biodiversity: Habitats in Decline, Species at Risk and Conservation.

12. Adaptation to Climate Change: Strategies for Reducing Vulnerabilities.

13. Mitigation: Reducing Emissions and Promoting Carbon Neutrality.

14. Circular Economy and Conscious Consumption: Reduction of the Environmental Impact.

15. Technological Innovation for a Sustainable Future: Clean and Scalable Technologies.

Chapter 1:
Introduction to Climate Change: Key Concepts and Global Context

Climate change represents one of the most pressing challenges of the modern era, with profound implications for the environment, society and the economy. In this chapter, we will explore the fundamental concepts related to climate change and lay the foundations for an in-depth understanding of its impacts and the actions needed to address it.

Climate Change Definition: Climate change refers to the long-term alteration of climate patterns on Earth. This change is mainly triggered by the increase in greenhouse gas emissions resulting from human activities, such as the use of fossil fuels, deforestation and industry.

Role of Greenhouse Gases: Greenhouse gases, such as carbon dioxide (CO2), methane (CH4) and nitrous oxide (N2O), trap heat in the earth's atmosphere through the greenhouse effect. This natural phenomenon is crucial for maintaining a temperature suitable for life, but human activities are intensifying the greenhouse effect, leading to an increase in global temperatures.

Global Temperature Rising: The planet's average temperature is steadily increasing, causing significant impacts such as melting polar ice, rising sea levels and altering regional climate regimes.

Scientific Evidence: Scientists study climate change by analyzing historical data, geological records, and climate models. This evidence shows that the current temperature rise is out of the ordinary compared to natural fluctuations.

Multidimensional Consequences: Climate change has knock-on effects on various aspects of life on Earth. This includes biodiversity loss, dwindling water resources, increased extreme weather events and risk to food security.

Global Context: Climate change is a transboundary challenge that requires global collaborative efforts. The 2015 Paris Agreement is a significant step in this direction, aiming to limit the global temperature increase to 1.5°C above pre-industrial levels.

Urgency for Action: Tackling climate change requires immediate and coordinated action. Solutions include reducing emissions, adopting renewable energy, promoting energy efficiency and adopting sustainable agricultural practices.

Chapter 2:
How Climate Works: From Solar Radiation to Atmospheric Dynamics

To understand the complex workings of the Earth's climate, it is essential to examine the processes that drive the transfer of energy in the atmosphere. This chapter explores the role of solar radiation and the atmospheric dynamics that shape our climate models.

Solar Radiation and Absorption: The sun is the primary source of energy for our planet. Solar radiation reaches the Earth in the form of electromagnetic waves, but part of it is absorbed by the atmosphere, the seas and the earth's surface.

Terrestrial Radiation: The heated Earth emits infrared radiation into space.

This radiation is absorbed by the greenhouse gases present in the atmosphere, such as CO2 and CH4, which re-emit it towards the surface, contributing to the warming of the atmosphere itself.

Atmospheric Circulation: Solar energy heats different regions of the Earth. Warm, light air rises, creating areas of low pressure, while cold, dense air descends, creating areas of high pressure. This causes winds and atmospheric currents, which affect the global climate.

Ocean Currents: Thermal energy absorbed by the oceans creates ocean currents that have a significant impact on climate. For example, the Gulf Stream carries heat to Europe, mitigating winter temperatures.

Coriolis Effect: Due to the Earth's rotation, winds and ocean currents are deflected in their trajectory. This phenomenon known as the "Coriolis effect" contributes to the formation of atmospheric convection loops and ocean currents.

Circulation Cells: Atmospheric currents organize into three main circulation cells in each hemisphere: the Hadley Cell, the Ferrel Cell and the Polar Cell. These vertical and horizontal motions are critical to the distribution of heat around the globe.

El Niño and La Niña: Oceanic phenomena known as El Niño and La Niña affect global climate by altering Pacific Ocean temperatures. These events can lead to climatic anomalies, such as droughts and heavy rains, in different parts of the world.

Climatic Ecosystems: The different climatic zones, such as tropical, temperate and polar regions, are determined by atmospheric dynamics. These ecosystems adapt to the prevailing climatic conditions, influencing the flora, fauna and natural resources.

In summary, the functioning of the climate is the result of an intricate ballet between solar radiation, atmosphere, oceans and land.

Understanding these dynamics is essential to address climate change and to develop adaptation and mitigation strategies.

Chapter 3:
The Scientific Evidence of Climate Change: Data, Models and Trends

Accumulating scientific evidence is critical to proving the existence and extent of climate change. In this chapter, we will explore how collected data, climate models, and observed trends confirm ongoing change and provide clear insight into global impacts.

Historical Records: Scientists analyze historical records of temperatures, precipitation and other climate parameters to identify changes over the centuries. These historical data reveal a net increase in temperatures over the past few decades.

Monitoring Networks: Global networks of weather stations, ocean buoys and satellites collect real-time data on temperature, sea level, greenhouse gas concentrations and other indicators. This data helps monitor current climate variations.

Climate Anomalies: Climate anomalies, such as sustained heat waves, melting ice, and intensification of extreme weather events, are consistent with patterns of climate change. These events highlight an evolving picture.

Climate Models: Scientists use complex climate models based on mathematical equations to simulate the behavior of the atmosphere, ocean and land. These models are able to reproduce past trends and predict possible future scenarios.

Warming Trends: Global temperatures are steadily rising, with noticeable increases in recent decades.

The Arctic is warming faster than other regions, causing melting ice and rising sea levels.

Ecosystem Change: The impacts of climate change are evident in ecosystems around the world. Coral reef bleaching events, coastal erosion and species migration are just a few examples of the changes taking place.

Scientific Agreement: The vast majority of the scientific community agrees that current climate change is primarily caused by human activity. Organizations such as the IPCC (Intergovernmental Panel on Climate Change) have stressed the urgency for action.

Past Comparison: Scientific evidence reveals that the current rate of climate change is exceptional compared to past natural fluctuations. This suggests a direct link to human activities and increased greenhouse gas emissions.

Future Scenario: Using advanced climate models, scientists project possible future scenarios in the absence of action to mitigate climate change. These projections point to a further increase in temperatures and impacts unless significant measures are taken.

In conclusion, scientific evidence gathered from a wide range of sources convincingly demonstrates that climate change is an undeniable reality. Data analysis, climate modeling and observed trends create a comprehensive picture of climate evolution and the challenges we face.

Chapter 4:
Impacts of Global Warming: Key Areas of Concern

Global warming is causing significant impacts on different parts of the world, putting ecosystems, economies and communities at risk. In this chapter, we will explore some of the key areas of concern where the effects of climate change are particularly evident.

1. Melting Polar Ice: The Arctic and Antarctica are experiencing rapid melting of ice due to rising temperatures. This is contributing to sea level rise, threatening coastal communities and biodiversity.

2. Sea Level Rise: Global warming is melting land ice and increasing the volume of the oceans. This causes sea levels to rise, with consequences for coastal areas, drainage systems and marine habitats.

3. Extreme Weather Events: Rising temperatures favor the intensification of extreme weather events such as hurricanes, cyclones, floods and droughts. These events can cause economic damage, loss of human life and alterations to the environment.

4. Biodiversity Loss: Climate-sensitive ecosystems, such as coral reefs and tropical forests, are experiencing increasing stress. This leads to the loss of biodiversity, threatening unique animal and plant species.

5. Poor Access to Water: Altered precipitation patterns can lead to prolonged droughts or sudden heavy rains. These changes affect access to drinking water, agriculture and water resource management.

6. Food Security: Climate change can affect food production through changes in temperature, rainfall and crop efficiency. This has implications for global food security.

7. Migration and Dislocation: Coastal communities and those dependent on vulnerable ecosystems are forced to migrate due to climate change. These movements can create social and economic challenges.

8. Human Health: Rising temperatures can lead to heat-related illnesses and the spread of vector-borne diseases such as mosquitoes and rats. Extreme weather events can also threaten the mental health of those affected.

9. Sustainability of Natural Resources: Climate change threatens the availability of resources such as wood, fish and fresh water. This can lead to conflicts and tensions between communities sharing these resources.

10. Infrastructure and Urban Assets: Infrastructure and urban assets not designed to withstand extreme weather events are exposed to risks of structural damage, flooding and disruption of essential supplies.
In these key areas of concern, the impact of global warming is already visible and requires concrete actions to mitigate the damage and adapt to future change.

Chapter 5:
Changes in Climate Regimes: Extreme Events, Vulnerable Areas and Migration

In the context of the current climate change scenario, the key word becomes "resilience". In this chapter, we will explore how to meet the challenges of climate change through innovation, adaptation and a hopeful perspective.

Facing Reality: Global warming is an undeniable reality, but resilience involves acknowledging the challenge without succumbing to fear. Understanding and acceptance are the first steps towards action.

Technological Innovation: Technological solutions play a crucial role in building resilience. From renewable energy to emissions capture and storage technologies, innovation can lead the transformation towards a low-carbon society.

Adaptation and Flexibility: Resilience requires the ability to adapt to ongoing changes. This may involve urban planning that takes account of extreme weather events or adopting agricultural practices that are more resilient to changes in weather conditions.

Resilient Communities: Local communities play a key role in building resilience. Actively involving people, educating and sharing resources can help mitigate impacts and develop support networks.

Holistic Approach: Resilience requires a holistic approach that considers the interconnections between the environment, the economy and society. The balance between these aspects is crucial to addressing the complex challenges of climate change.

Green Economy: Promoting a green economy based on principles of sustainability and emission reduction is an integral part of resilience. This implies the transition from carbon-intensive sectors to more sustainable sectors.

Global Collaboration: Resilience knows no boundaries. International collaboration is essential to address climate change. The Paris Agreement represents a step forward, but further efforts and commitments are needed.

Hope and Action: Resilience requires trust that we can meet today's challenges. Every step, big or small, towards sustainability and adaptation contributes to building a better future.

Education and Awareness: Raising awareness of climate change and resilience strategies is essential. Education can inspire individual and collective action towards more sustainable choices.

Looking Forward: Despite ongoing challenges, resilience drives us to look forward with hope. Actions taken today can lay the foundations for a more resilient, sustainable and inclusive world.
In a fast-changing world, resilience invites us to turn challenges into opportunities, driven by innovation and a belief in the power of human action.

Chapter 6:
Oceans in Danger: Acidification, Sea Level Rise and Marine Biodiversity

The oceans, vital pillars of our planet, are facing increasing threats from climate change. In this chapter, we will explore the challenges related to ocean acidification, sea level rise and marine biodiversity loss.

Ocean Acidification: Rising CO2 emissions are causing a buildup of carbon dioxide in the oceans. This induces a process known as acidification, which makes the seas more acidic. This phenomenon can damage corals, molluscs and other organisms with calcium carbonate shells.

Impact on Marine Life: Ocean acidification can negatively affect marine life at various levels of the food chain. Organisms such as corals, plankton, molluscs and fish may find it difficult to survive and grow in more acidic waters.

Sea Level Rise: Global warming is causing land ice to melt, contributing to sea level rise. This poses a threat to coastal communities, marine habitats and wetlands.

Coastal Erosion: Rising sea levels can lead to coastal erosion, with land loss and damage to coastal infrastructure. This puts millions of people living in coastal areas at risk.

Loss of Habitat and Biodiversity: Rising sea levels can lead to the loss of coastal habitats such as mangroves and seagrass beds. These ecosystems are essential for marine biodiversity and storm protection.

Risk to Coastal Communities: Coastal communities are particularly vulnerable to sea level rise and extreme weather events. These populations face challenges such as migration, food security and adaptation to new conditions.

Chapter 7:
The Role of Human Activities: Greenhouse Gas Emissions, Deforestation and Pollution

Human activities play a central role in accelerating climate change and altering environmental balances. In this chapter, we will examine the contribution of greenhouse gas emissions, deforestation and pollution to the current climate crisis.

Greenhouse Gas Emissions**: The use of fossil fuels such as coal, oil and natural gas is primarily responsible for greenhouse gas emissions such as CO2 and methane. These gases trap heat in the atmosphere, causing global warming.

Deforestation**: The clearing of forests at an alarming rate has devastating consequences. Forests absorb CO2 and act as "carbon sinks". Deforestation releases this carbon into the atmosphere, increasing greenhouse gas concentrations.

Air Pollution: Air pollution from industry, transportation and other industrial sources not only compromises air quality, but can also affect climate. Fine particles and aerosols can reflect or absorb sunlight, changing the thermal dynamics of the atmosphere.

Intensive Agriculture: Excessive use of fertilizers and inefficient management of agricultural waste can lead to emissions of nitrous oxide, a potent greenhouse gas. Livestock farming also produces methane through animal digestion and waste decomposition.

Biodiversity Impacts: Human activities such as deforestation and pollution threaten biodiversity. The loss of habitat and the introduction of polluting substances can put animal and plant species at risk, compromising the ecological balance.

Possible Solutions: Reducing greenhouse gas emissions requires a transition to clean energy sources, improving energy efficiency and promoting sustainable agricultural practices. Forest conservation and reforestation are essential for carbon capture.

Role of the Individual: Individual choices, such as the responsible use of energy, the adoption of sustainable consumption habits and participation in awareness-raising initiatives, can help reduce the impact of human activities on the climate .

Laws and Regulations: Governments can play a crucial role in adopting policies and regulations that encourage the reduction of emissions and the protection of the environment. International agreements such as the Paris Agreement are examples of joint efforts to address the climate crisis.

In summary, human involvement is at the heart of climate change. Understanding and reducing the role of human activities in greenhouse gas emissions, deforestation and pollution is essential to mitigate impacts and build a more sustainable future.

Chapter 8:
Energy and the Green Transition: From Fossil Fuel Dependency to Renewable Sources

Energy is at the heart of the modern economy and society, but its production has had a significant impact on the climate. In this chapter, we will explore the need to address dependence on fossil fuels through the transition to renewable energy sources.

Fossil Fuels and Climate Change: The consumption of coal, oil and natural gas for energy production is the main driver of greenhouse gas emissions. These fuels contribute to the greenhouse effect and global warming.

Renewable Energy Sources: Renewable energy sources, such as sun, wind, water and biomass, are inexhaustible and produce little or no greenhouse gas emissions. These sources offer a sustainable alternative to fossil fuels..

Solar Energy: Solar energy uses sunlight to generate electricity. Photovoltaic and thermal solar panels are increasingly used to power homes, businesses and industrial plants.

Wind Energy: Wind farms capture the kinetic energy of the wind and transform it into electricity. Onshore and offshore wind turbines are becoming key components of renewable energy production.

Hydroelectric Energy: Dams and hydroelectric plants use the energy of moving water to generate electricity. This source is already widely used in many parts of the world.

Biomass and Biogas: Biomass, such as wood and organic waste, can be converted into energy through combustion processes or biogas production. This helps reduce dependence on fossil fuels.

Challenges and Obstacles: The transition to renewable energy sources brings challenges such as the intermittence of resources (sun and wind are not constant), the need for adequate infrastructure and the training of a skilled workforce.

Investments and Incentives: Investments in renewable energy are crucial to accelerate the transition. Governments, companies and institutions are providing financial incentives and support policies to promote the transition to clean sources.

Economic and Social Impact: The green transition can create new jobs, stimulate technological innovation and reduce public health costs due to air pollution.

Role of the Private Sector: Companies can play a key role in adopting sustainable practices and investing in renewable energy, contributing to the reduction of emissions and the adoption of cleaner solutions.

The transition to renewable energy sources is a crucial step in the fight against climate change. Moving away from dependence on fossil fuels in favor of a sustainable energy supply is key to building a low-carbon future.

Chapter 9:
Sustainable Agriculture and Food: Climate Impacts and Agroecological Solutions

Agriculture is both affected by and contributing to climate change. In this chapter, we will explore climate impacts on agriculture, food security risks, and agroecological solutions for a more sustainable food system.

Climate Impacts on Agriculture: Climate change can alter precipitation patterns, cause droughts, heat waves and increase the frequency of extreme weather events. These impacts affect agricultural production, crop yields and soil quality.

Food Security Risk: Climate impacts on agriculture can threaten global food security. Reductions in the production of key crops can lead to food shortages and rising prices.

Agroecological Solutions: The agroecological approach promotes sustainable agricultural practices that take into account the environment, biodiversity and local needs. These practices include crop diversification, conscious use of water, and reducing the use of chemical fertilizers.

Precision Farming: Technology can help optimize the use of agricultural resources. Drones, sensors, and data analytics can drive better field management and irrigation decisions.

Climate Resilient Agriculture: Growing crop varieties that are resilient to local climate conditions can help mitigate impacts. These varieties can be selected for their tolerance to drought or high temperatures.

Reducing Food Waste: Reducing food waste is an integral part of a sustainable food system. This not only conserves resources but also reduces greenhouse gas emissions generated by the production of food that is not consumed.

Sustainable Farming of Animals: Intensive breeding is responsible for a significant part of greenhouse gas emissions. Approaches such as free range, rotary grazing and sustainable feeding can reduce environmental impact.

Role of Innovation: Research and innovation in agriculture can lead to more climate resilient crops, more efficient farming practices and conservation methods that minimize waste.

Community Involvement: Involving local communities in agriculture and food production can promote practices that are sustainable and adapted to the specific needs of each region.

Developing a sustainable food system is crucial to addressing the impacts of climate change and ensuring food security for future generations. The adoption of agroecological solutions and innovation in the agricultural sector can contribute to a future where food production is in harmony with the environment.

Chapter 10:
Resilient Cities: Urban Planning, Transport and Pollution Reduction

Cities, which are home to the majority of the world's population, play a crucial role in the fight against climate change. In this chapter, we'll explore the importance of urban planning, sustainable transportation, and pollution reduction to create more resilient cities.

Sustainable Urban Planning: City planning must take into account the impact of climate change. The creation of green spaces, the adoption of eco-friendly materials and energy efficiency in buildings are key aspects.

Sustainable Transport: Cities generate a large part of greenhouse gas emissions due to transport. Investing in efficient public transport, promoting cycling and the adoption of electric vehicles can reduce emissions and improve air quality.

Shared mobility: Car sharing and ride sharing services can reduce the number of vehicles on the road, reducing congestion and emissions.

Reducing Air Pollution: Reducing urban air pollution not only improves the health of citizens, but also reduces the overall impact on the climate. Tighter regulations on the emission of pollutants may contribute to this cause.

Sustainable Buildings: Designing sustainable buildings with low-carbon materials and energy and cooling solutions can reduce energy consumption and emissions.

Risks from Extreme Weather Events: Cities are often exposed to extreme weather events such as floods, storms and heat waves. Urban planning should include defense and adaptation infrastructure to address these risks.

Technological Innovation: The use of smart technologies, such as data monitoring and advanced energy management, can help make cities more energy efficient and resilient.

Community Involvement: Involving citizens in decisions about urban planning and environmental solutions can promote greater awareness and active participation.

Positive Side Effects: Building resilient cities not only tackles climate change, but can also improve citizens' quality of life, create sustainable jobs and increase social cohesion.

Building resilient cities requires integrated efforts of planning, technology, policy and active participation. Cities can play a vital role in tackling climate change and shaping a more sustainable future for generations to come.

Chapter 11:
Threatened Biodiversity: Habitats in Decline, Species at Risk and Conservation

Biodiversity loss is one of the most pressing challenges of our time, with profound consequences for global ecosystems. In this chapter, we will explore the threat to biodiversity, endangered species and conservation strategies.

Natural Habitat Decline: Land transformation for agricultural, urban and industrial purposes has led to the destruction of natural habitats such as forests, wetlands and grasslands. This loss reduces the availability of living spaces for many species.

Extinction of Species: Climate change, illegal hunting, deforestation and other human activities have led to the extinction of numerous plant and animal species. Once a species goes extinct, the entire ecosystem can be disrupted.

Endangered Species: Many species are at risk of extinction or are considered vulnerable. These include iconic animals such as tigers, elephants and rhinos, but many lesser-known species also play crucial roles in ecosystems.

Corals and Coral Reefs: Marine ecosystems, such as coral reefs, are threatened by ocean acidification, pollution and global warming. These habitats are home to a variety of marine species and play a key role in marine biodiversity.

Conservation Strategies: The creation of protected areas, the restoration of natural habitats and the fight against illegal hunting are some of the strategies for conservation. Involvement of local communities is often crucial to the success of these initiatives.

In Situ and Ex Situ Conservation: In situ conservation focuses on the protection of species in their natural habitat, while ex situ conservation involves the preservation of species in captivity or in botanic gardens.

Role of Nature Reserves: Nature reserves, such as national parks and marine protected areas, play a key role in the conservation of biodiversity, providing refuge for many threatened species.

Collaborative Conservation: International cooperation is essential for the conservation of species that migrate across national boundaries. Bilateral and multilateral agreements can facilitate the monitoring and protection of these species.

Education and Awareness: Public awareness and education about the value of biodiversity can stimulate greater commitment to its conservation.

Preserving biodiversity is essential for ecosystem health, food security and global sustainability. Adopting purposeful conservation strategies and promoting sustainable practices are key to protecting the amazing variety of life that shares our planet.

Chapter 12:
Adapting to Climate Change: Strategies for Reducing Vulnerabilities

Adapting to climate change is essential to addressing the inevitable impacts we are already experiencing. In this chapter, we will explore key strategies to reduce vulnerabilities and improve capacity to meet the challenges of climate change.

Climate Risk Assessment: Understanding the specific climate threats a region faces is the first step. Climate risk assessments guide the identification of the most vulnerable areas and adaptation needs.

Resilient Spatial Planning: Urban and rural planning should integrate climate impacts. This may include creating flood-protected areas, avoiding construction in high-risk areas, and promoting resilient buildings.

Water Management: Planning for efficient water use, managing water resources, and creating improved drainage infrastructure are key to addressing the challenges associated with precipitation variations.

Agriculture and Food Security: Sustainable agricultural practices, such as growing drought-tolerant crops and adopting soil conservation techniques, can help address climate impacts on food production.

Early Warning Systems: Developing early warning systems for extreme weather events, such as storms and floods, can save lives and help communities prepare and respond in a timely manner.

Reducing Poverty and Inequalities: The most vulnerable communities are often the most affected by climate change. Reducing poverty, improving access to basic services and addressing inequalities can increase resilience.

Adaptive Technologies: Adopting technologies such as efficient irrigation systems, climate-resilient building materials, and innovative cultivation technologies can help mitigate negative impacts.

Ecosystem Conservation: Natural ecosystems such as forests, wetlands and coral reefs can act as buffers against the impacts of climate change. The conservation of these ecosystems is essential for resilience.

Public Health Planning: Adaptation also includes preparing for public health impacts, such as the increase in heat-related illnesses or the spread of vector-borne diseases.

Community Engagement: Involving local communities in adaptation decisions and implementation of strategies is crucial to ensure that solutions are tailored to local needs and knowledge.

Chapter 13:
Mitigation: Reducing Emissions and Promoting Carbon Neutrality

Mitigation is a key component in the fight against climate change, aiming to reduce greenhouse gas emissions and achieve carbon neutrality. In this chapter, we will explore strategies and approaches to mitigating the impacts of climate change.

Reducing Greenhouse Gas Emissions: Reducing emissions is the core of mitigation. This can be achieved through the adoption of low carbon technologies, energy efficiency and the transition to renewable energy sources.

Renewable Energy: The transition from fossil energy sources to renewable sources, such as solar and wind, is essential to reduce CO2 emissions related to energy production.

Energy Efficiency: Improving efficiency in energy production and use can significantly reduce emissions. This can include the adoption of more efficient technologies and conscious consumption practices.

Sustainable Mobility: Promoting the use of public transport, bicycles, electric vehicles and shared mobility services can reduce the impact of emissions from the transport sector.

Energy Savings in Buildings: Improving the energy efficiency of buildings through thermal insulation, the use of energy efficient windows and the adoption of efficient heating and cooling systems can reduce energy consumption .

Sustainable Agriculture: Sustainable agricultural practices, such as soil management, reducing the use of fertilizers and controlling emissions from agricultural waste, can help reduce greenhouse gas emissions from the agricultural sector.

Waste Management: Sustainable waste management, through recycling, composting and reducing waste production, can avoid the emission of methane, a powerful greenhouse gas.

Carbon Neutrality: Carbon neutrality, or carbon neutrality, is achieved when the greenhouse gas emissions produced are equal to the emissions that are removed from the atmosphere through actions such as reforestation and the capture and storage of emissions.

Carbon Market: Carbon trading schemes create economic incentives for companies to reduce their emissions, encouraging the transition to low-carbon practices.

Global Commitment: Mitigation efforts require a coordinated global commitment. International agreements such as the Paris Agreement set emission reduction targets for participating countries.

Mitigation requires multi-sector efforts and cooperation between governments, industries, communities and individuals. Reducing emissions and pursuing carbon neutrality is key to limiting global temperature rise and protecting our planet for future generations.

Chapter 14:
Circular Economy and Conscious Consumption: Reduction of the Environmental Impact

The adoption of a circular economy model and conscious consumption are essential for reducing the environmental impact of human activities. In this chapter, we will explore how these approaches can contribute to sustainability and the fight against climate change.

Circular Economy: The circular economy is based on the principle of reducing, reusing and recycling resources to avoid waste. This includes designing durable products, recovering materials and reducing waste generation.

Reuse and Recycle: Promoting the reuse and recycling of materials reduces the need to extract new resources, reduces the production of waste and contributes to a more efficient use of raw materials.

Circular Design: Designing products with a circular vision in mind, considering their complete life cycle and facilitating the recovery of materials after use, is an important step in reducing waste.

Conscious Consumption: Being aware of consumption choices can reduce environmental impact. Opting for durable products, avoiding the excessive use of packaging and looking for products with a low carbon footprint are examples of conscious consumption.

Sustainability in the Supply Chain: Companies can promote sustainable practices throughout their supply chain, demanding sustainable raw materials, ethical practices and waste reduction.

Sharing Economy: Sharing-based business models, such as car-sharing services, clothing rentals, and tool libraries, can reduce the need to excessively own assets.

Reducing Food Footprint: Reducing meat consumption, avoiding food waste and preferring local and seasonal products can help reduce the environmental impact associated with agriculture and food production.

Promoting Circularity: Governments can promote the adoption of circular practices through tax incentives, waste management laws and awareness campaigns.

Education and Awareness: Education on sustainable consumption practices and the importance of the circular economy can increase awareness and adoption of responsible behavior.

Economic Benefits: The circular economy can lead to long-term economic savings, stimulate innovation and create new business opportunities.

The adoption of circular economy practices and conscious consumption is crucial to reduce the excessive use of resources and the environmental impact of human activities. These approaches can significantly contribute to environmental sustainability and the reduction of greenhouse gas emissions.

Chapter 15: Technological Innovation for a Sustainable Future: Clean and Scalable Technologies

Technological innovation plays a crucial role in tackling climate change and creating a sustainable future. In this chapter, we will explore some of the key technologies that can help reduce emissions and promote sustainability.

Energy from Renewable Sources: Technologies such as photovoltaic solar panels, wind turbines and hydroelectric power plants are essential for producing clean energy and reducing greenhouse gas emissions.

Advanced Batteries: High capacity batteries are essential for storing energy from renewable sources such as sun and wind, allowing for constant release even when the sun is not shining or the wind is not blowing.

Carbon Capture and Storage Technologies: These technologies make it possible to capture CO2 emissions produced by industrial or energy sources and store them safely, thus reducing the environmental impact.

Electric Mobility: Electric and hybrid vehicles reduce greenhouse gas emissions from the transport sector. Additionally, the growth of charging infrastructure is contributing to greater adoption.

Energy Efficient Technologies: Smart sensors, energy management systems and energy-efficient lighting technologies are key tools to reduce energy consumption in buildings and industry.

Precision Agriculture: The use of drones, sensors and data analysis can optimize the use of agricultural resources, reducing the use of water and fertilizers and improving crop yields.

Blockchain for Traceability: Blockchain technology can be used to trace the supply chain and ensure the sustainable sourcing of products.

Water Technologies: Innovative solutions for seawater desalination, wastewater treatment and water resource monitoring contribute to more sustainable water management.

Green Building: Eco-friendly building materials, efficient heating and cooling systems, and waste management technologies in construction promote sustainable building.

Ocean Technologies: Wave, tidal and thermal energy from the oceans can be harnessed for energy production, contributing to a transition towards renewable sources.

Technological innovation offers promising solutions to tackle climate change and promote sustainable development. The adoption of clean and scalable technologies is crucial to reducing emissions and creating a future where the environment and the economy can coexist in harmony.

Chapter 16:
Global Climate Policies: Paris Agreement, COP and International Efforts

International cooperation is key to addressing climate change on a global scale. In this chapter, we will explore key global policies, agreements and international efforts aimed at mitigating climate impacts.

Paris Agreement: Adopted in 2015 under the United Nations Framework Convention on Climate Change (UNFCCC), the Paris Agreement sets out the goal of limiting the increase in global mean temperature to "well below by 2°C above pre-industrial levels, with efforts to limit the increase to 1.5°C. Member States commit to presenting National Emissions Reduction Plans (NDCs) and regularly reviewing them in a more ambitious way.

Conferences of the Parties (COP): COPs are annual meetings of signatory countries to the Paris Agreement to discuss and negotiate climate action strategies.

COP21 2015 led to the Paris Agreement. These conferences serve as a platform for discussions, sharing best practices and setting common goals.

Emissions Reduction Plans (NDCs): National Emissions Reduction Plans are voluntary commitments by countries to reduce their greenhouse gas emissions. These plans reflect the actions each country intends to take to contribute to the goals of the Paris Agreement.

Climate Finance: International agreements provide for financial support to developing countries to tackle climate change and adopt sustainable practices. This funding is being made available through various sources, including developed countries, international financial institutions and dedicated climate funds.

Kyoto Protocol: Adopted in 1997, the Kyoto Protocol was the first global agreement to reduce greenhouse gas emissions. Although it is no longer in force, it has set an important precedent for international cooperation on the climate issue.

Cities Network: In addition to government efforts, many cities around the world have adopted plans and policies to reduce emissions and increase resilience to climate change, collaborating through networks such as the "Covenant of Mayors" and "C40 Cities ".

International Treaty and Coordination: Combating climate change requires continued global collaboration, coordinating efforts at the international and individual levels. Agreements and joint efforts help create coherent action and share best practices.

Challenging Global Emissions: Tackling climate change requires the involvement of all countries, regardless of the size of their economies. A common challenge is to reduce global emissions and adopt more sustainable behaviors.

Global climate policies aim to create a collective commitment to tackle climate change and protect the environment. International collaboration is key to mitigating climate impacts and building a sustainable future for generations to come.

Chapter 17:
Community Involvement: Local Initiatives and Public Participation

Community engagement is essential to addressing climate change effectively. In this chapter, we will explore how local initiatives and public participation can help create a positive and lasting impact.

Awareness Raising and Education: Informing communities about the risks and impacts of climate change is the first step. Education can foster a deeper understanding of the challenge and inspire concrete action.

Participatory Planning: Involving citizens in planning adaptation and mitigation strategies can ensure that solutions respond to local needs and realities.

Promotion of Sustainable Practices: Local initiatives can encourage the adoption of sustainable behaviors, such as recycling, the use of public transport and saving energy.

Community Gardening and Urban Agriculture: Creating shared green spaces and promoting urban agriculture not only improves the environment, but can also strengthen social cohesion and food security.

Community Energy Projects: Communities can work together to develop renewable energy projects, such as shared solar panels, that reduce emissions and provide clean energy.

Environmental Monitoring: Involving citizens in environmental monitoring can provide valuable data on air pollution, water quality and other environmental issues.

Awareness Campaigns: Creating local awareness campaigns can inspire action and active participation, highlighting the importance of individual and collective change.

Co-creation of Solutions: Engaging communities in co-creation of solutions can lead to innovative ideas and approaches better suited to local needs.

Sustainability Networks: The creation of local sustainability networks allows communities to share resources, experiences and ideas, amplifying the impact of initiatives.

Role of Local Organizations: Local non-profit organizations, educational institutions and businesses can play a key role in leading action at the community level.

Engaging communities is key to creating a widespread change movement. Local initiatives can have a significant impact on tackling climate change by promoting sustainable practices and stimulating innovation.

Chapter 18:
Education and Awareness: Communicating the Urgency of Climate Change

Communicating the urgency of climate change is key to mobilizing action at all levels of society. In this chapter, we will explore the importance of education and awareness in tackling climate change.

Science-Based Information: Communicating science-based data and information helps provide an accurate understanding of climate change and its impacts.

Clear and Accessible Communication: Using clear and accessible language makes information on climate change accessible to a wider audience, including people of different ages and backgrounds.

Engaging Emotions: Personal stories and testimonials can spark empathy and emotional engagement, inspiring people to take concrete action.

Visual Communication: The use of graphs, infographics and visualizations can simplify complex concepts and make climate change trends clearer.

Climate Change's Link to Everyday Life: Showing how climate change affects everyday life, from food supply to the climate, can make the issue more relevant to people.

Community Effect: Engaging with local communities, schools, workplaces and interest groups can create a sense of ownership of a shared cause.

Youth Participation: Students and young people can be influential agents of change. Climate change education in schools can encourage active participation.

Providing Concrete Solutions: Communicate practical solutions and actions people can take, from saving energy to reducing the use of single-use plastics.

Through Different Channels: Use social media, community events, conferences and traditional media to reach different sections of society.

Focus on the Future: Showing how today's actions will affect the future of the next generations can increase the sense of responsibility.

Education and awareness are powerful tools to promote understanding of climate change and prompt action. Effectively communicating the urgency of the challenge and the opportunities for change can inspire individuals, communities and governments to make sustainable choices for a better future.

Chapter 19: Building a Sustainable Future: Individual and Collective Actions for a Better Planet

The path to a sustainable future requires the commitment of each of us, both individually and collectively. In this chapter, we'll explore actionable actions that can help build a healthier, more balanced planet.

1. Reducing Carbon Emissions: Use public transport, bike or walk whenever possible. Reduce excessive heating and cooling in your home and replace light bulbs with energy-efficient LEDs.

2. Reducing Plastic Use: Limit the use of single-use plastics, such as water bottles and grocery bags. Opt for reusable products and prefer sustainable materials.

3. Sustainable Food Choices: Reduce your consumption of meat and dairy products, which are resource-intensive, and opt for a plant-based diet. Prefer local and organic products.

4. Reduce Food Waste: Plan meals, store leftovers, and use ingredients creatively to reduce food waste.

5. Save Water: Fix water leaks, reduce time in the shower and use water responsibly, both indoors and outdoors.

6. Support Renewable Energy: Choose energy suppliers that use renewable sources such as the sun or wind. If possible, install solar panels or adopt other green technologies.

7. Attend Awareness Event: Join marches, workshops and local initiatives that promote sustainability and spread awareness.

8. Engage Your Community: Organize cleanup projects, tree plantings, or other initiatives to engage your community in environmental action.

9. Reduce the Impact of Clothing: Go for quality and long-lasting clothes, opt for recycling and the thrift store and support sustainable clothing brands.

10. Education & Sharing: Share your knowledge about the environment and encourage others to join you in adopting sustainable habits.

11. Encourage Sustainable Policies: Participate in public debate, write to your political representatives and support policies that promote sustainability.

12. Sustainability at Work: Look for ways to reduce the environmental impact in your workplace, such as saving energy and recycling.

13. Reduce Your Digital Footprint: Reduce the energy consumption of electronic devices, limit the amount of useless emails and use the web responsibly.

14. Sustainability in Travel: Opt for sustainable travel, reduce the use of short-haul flights and look for low-impact accommodation and activities.

15. Support Conservation: Contribute to organizations and projects that are committed to environmental conservation and biodiversity protection.

Taking small daily actions and getting involved in collective initiatives can make a big difference in building a sustainable future. Every step towards sustainability brings us closer to a better planet for present and future generations.

Chapter 20:
Concluding Chapter: Towards a Sustainable Future

In this journey through the key concepts of climate change and the global context in which it fits, we have explored the far-reaching impact humanity is having on our planet. From analyzing how climate works and the scientific evidence of climate change, to the consequences of growing environmental impacts, we've tackled the data with candor and the hope of inspiring action.

It has become clear that climate change is not an abstract phenomenon or a challenge that belongs to the distant future. It is a current problem that requires an urgent response. Evidence of changes in our oceans, forests and ecosystems has underlined the need for collective and determined action.

Global policies such as the Paris Agreement and the COP conferences have shown that international cooperation is essential for tackling climate change. However, the real challenge lies in the actions of each individual, community, organization and nation. We explored actions we can take on an individual level, from reducing emissions to promoting sustainable choices in our daily lives.

Awareness raising, education and public participation are key to creating a culture of sustainability. We have seen how science-based information, communicated in an accessible and emotionally engaging way, can drive change. Engaging with communities, youth and local leaders is essential to ensure long-term commitment.

Tackling climate change will require sacrifice, adaptation and innovation. But it also offers an opportunity to build a more equitable, resilient and connected world. We have the power to shape a future where the balance between humanity and the environment is restored.

Each of us has a part to play in creating a sustainable future. It's time to move from words to action, to challenge the status quo and inspire others to follow suit. Together, we can meet the challenge of climate change and build a world where the environment thrives and future generations can enjoy the beauty and richness of our planet. Change is possible and necessary. The future is in our hands.